Everyday Mathematics®

Student Math Journal 1

The University of Chicago
School Mathematics Project

McGraw Hill Wright Group

The McGraw-Hill Companies

UCSMP Elementary Materials Component

Max Bell, Director

Authors

Max Bell
Jean Bell
John Bretzlauf*
Amy Dillard*
Robert Hartfield

Andy Isaacs*
James McBride, Director
Kathleen Pitvorec*
Peter Saecker

Technical Art

Diana Barrie*

Second Edition only

Photo Credits

Phil Martin/Photography, Jack Demuth/Photography, Cover Credits: Leaf, tomato, buttons/Bill Burlingham Photography, Tree rings background/Peter Samuels/Stone, Photo Collage: Herman Adler Design Group

Contributors

Robert Balfanz, Judith Busse, Ellen Dairyko, Lynn Evans, James Flanders, Dorothy Freedman, Nancy Guile Goodsell, Pam Guastafeste, Nancy Hanvey, Murray Hozinsky, Deborah Arron Leslie, Sue Lindsley, Mariana Mardrus, Carol Montag, Elizabeth Moore, Kate Morrison, William D. Pattison, Joan Pederson, Brenda Penix, June Ploen, Herb Price, Dannette Riehle, Ellen Ryan, Marie Schilling, Susan Sherrill, Patricia Smith, Robert Strang, Jaronda Strong, Kevin Sweeney, Sally Vongsathorn, Esther Weiss, Francine Williams, Michael Wilson, Izaak Wirzup

Permissions

page 187: *North American Indian Stickers* by Madeleine Orban-Szontagh. Dover Publications, Inc.

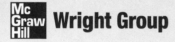

Copyright © 2004 by Wright Group/McGraw-Hill.

Send all inquiries to:
Wright Group/McGraw-Hill
P.O. Box 812960
Chicago, IL 60681

Printed in the United States of America.

ISBN 0-07-584441-9

14 15 16 DBH 10 09 08 07

Contents

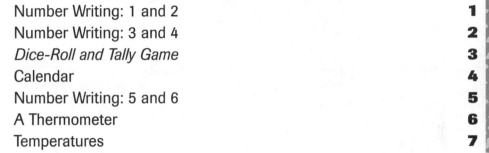

Unit 3: Visual Patterns, Number Patterns and Counting

Unit 4: Measurement and Basic Facts

Unit 5: Place Value, Number Stories, and Basic Facts

Activity Sheets

Number Writing: 1

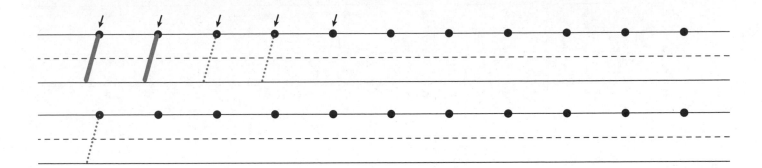

1	1 + 0	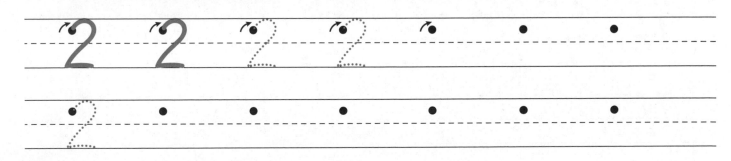
/	2 − 1	
uno		one

Draw a 1 picture.

Number Writing: 2

2	1 + 1	
//	3 − 1	
dos		two

Draw a 2 picture.

Number Writing: 3

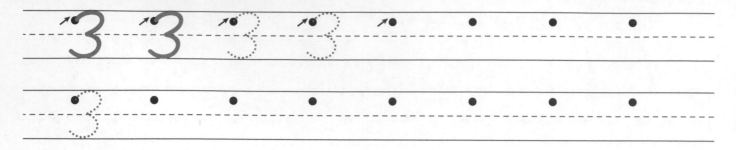

3	2 + 1	
///	4 − 1	
trés		three

Draw a 3 picture.

Number Writing: 4

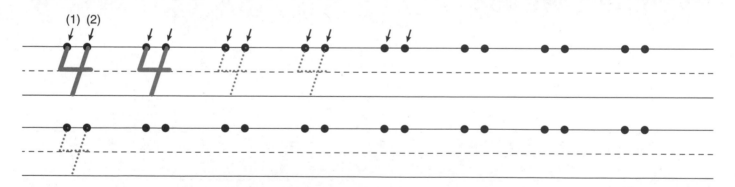

4	3 + 1	
////	5 − 1	
cuatro		four

Draw a 4 picture.

Use with Lesson 1.7.

Dice-Roll and Tally Game

1. Roll a die. Use tally marks to record the results on this chart.

	Tallies
1	
2	
3	
4	
5	
6	

2. Record the number of times each number came up.

———————

———————

———————

———————

———————

———————

Date _____

Month _____

	Sunday	Monday	Tuesday	Wednesday	Thursday	Friday	Saturday

Number Writing: 5

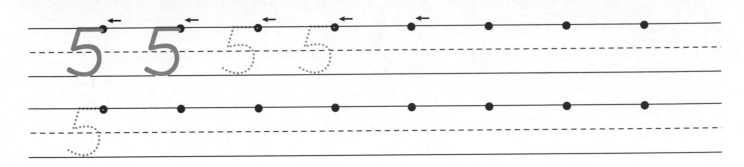

5	4 + 1	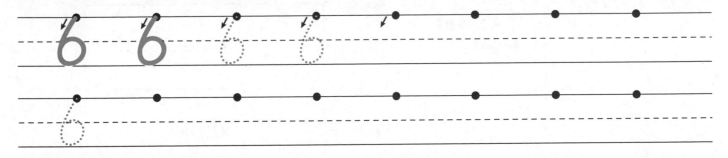
₪₪₪	6 − 1	
cinco		five

Draw a 5 picture.

Number Writing: 6

6	5 + 1	
₪₪₪ /	7 − 1	
seis		six

Draw a 6 picture.

A Thermometer

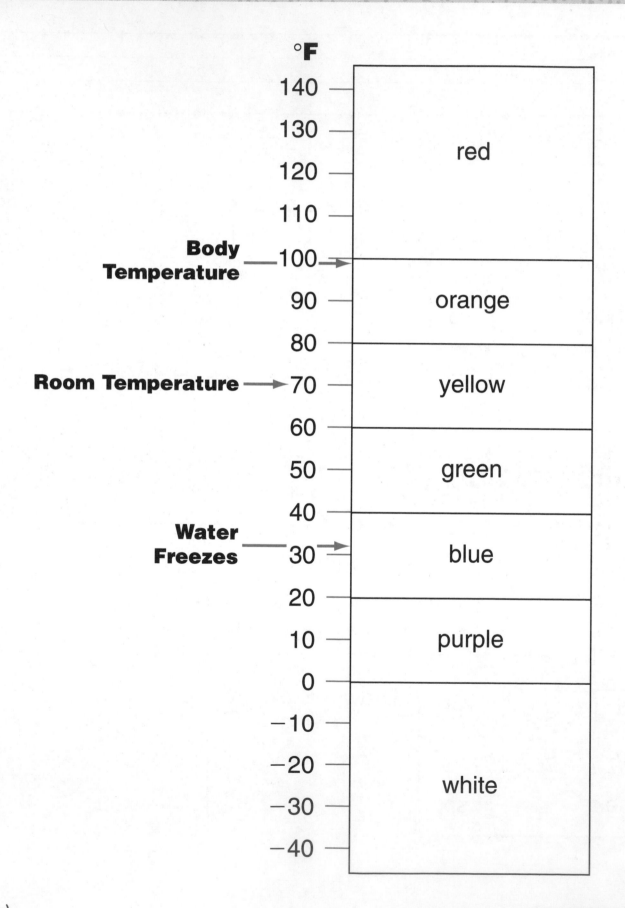

Use with Lesson 1.12.

Temperatures

Color each thermometer with the color for the temperature's zone.

Draw a picture of a person wearing the right kind of clothing for that temperature.

1.

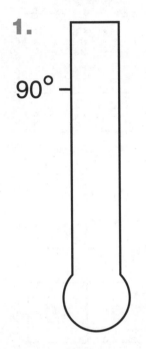

90°

2.

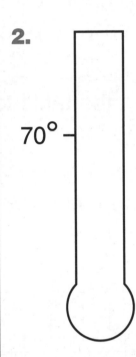

70°

3.

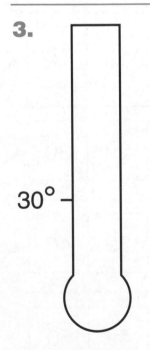

30°

4.

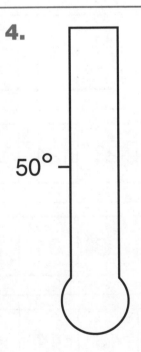

50°

Rolling for 50

Materials
- ❑ a die
- ❑ a marker for each player
- ❑ a gameboard for each player

Directions Take turns. Put your marker on START on your own gameboard.

Roll the die. Look in the table to see how many spaces to move.

Decide which number will be your FINISH number. The first player to reach or cross FINISH wins.

Roll	Spaces
1	3 up
2	2 back
3	5 up
4	6 back
5	8 up
6	10 up

START	1	2	3	4	5	6	7	8	9	10

11	12	13	14	15	16	17	18	19	20

21	22	23	24	25	26	27	28	29	30

31	32	33	34	35	36	37	38	39	40

41	42	43	44	45	46	47	48	49	50

Use with Lesson 2.1.

Information about Me

My first name is _____.

My second name is _____.

My last name is _____.

I am _____ years old.

Put candles on your cake.

My area code and
home telephone number are

(____ ____ ____) ____ ____ ____ – ____ ____ ____ ____

    (area code)   (telephone number)

Important Phone Numbers

Emergency number: ____ ____ ____

Police station number:

(____ ____ ____) ____ ____ ____ – ____ ____ ____ ____

Fire station number:

(____ ____ ____) ____ ____ ____ – ____ ____ ____ ____

Number Writing: 7

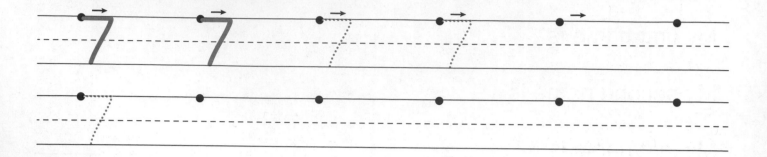

7	6 + 1	⚅⚁
	~~IIII~~ II	8 − 1
	siete	seven

Draw a 7 picture.

Number Writing: 8

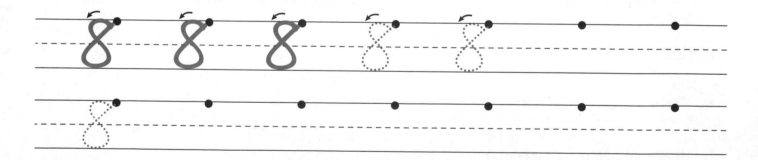

8	7 + 1	⚄⚂
	~~IIII~~ III	9 − 1
	ocho	eight

Draw an 8 picture.

Use with Lesson 2.2.

Math Boxes 2.3

1. Make a tally for 12.

2. Write the number.

_____   _____

3. Count up by 1s.

    7,   8,   9,

_____, _____, _____,

_____, _____, _____,

_____, _____

4. Write the number that is 1 more.

7 _____

15 _____

19 _____

Number Writing: 9

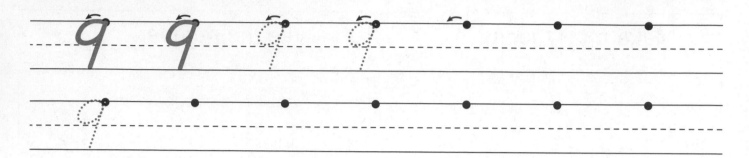

<table>
<tr><td>

9

8 + 1

⫲⫲⫲ ||||

nueve

</td><td>

🎲 🎲

10 − 1

nine

</td></tr>
</table>

Draw a 9 picture.

Number Writing: 0

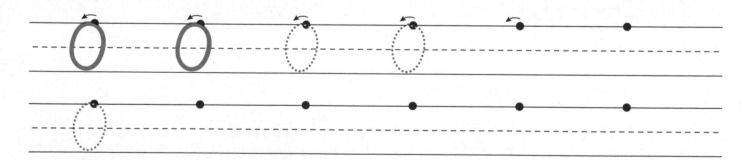

<table>
<tr><td>

0

0 + 0

1 − 1

cero

</td><td>

☐

zero

</td></tr>
</table>

Use with Lesson 2.4.

Math Boxes 2.4

1. Count up by 1s.

14,   *15,*   *16,*

———— , ———— , ———— ,

———— , ———— , ———— ,

———— , ————

2. Make a tally for 14.

3. Write the number that comes before.

———— 10

———— 15

———— 21

4. Circle the winner of *Top-It*.

11   15

Date

1. Circle the number that could be the mystery number.

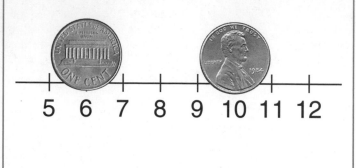

5   6   7   8   9 10 11 12

    5   9   12

2. How many?

HHT HHT HHT III

_____

3. Count up by 1s.

17,   18,   19,

_____, _____, _____,

_____, _____, _____,

_____, _____

4. Circle the winner of *Top-It*.

19	9

14 (fourteen)

Telling Time

1. Record the time.

_____ o'clock

_____ o'clock

_____ o'clock

_____ o'clock

2. Draw the hour hand.

2 o'clock

6 o'clock

Math Boxes 2.6

1. Write the number.

_____ _____

2. Write your phone number.

_____

3. Make sums of 10 pennies.

Left Hand	Right Hand
4	6
3	
	2

4. Write the next number.

7 _____

13 _____

25 _____

Use with Lesson 2.6.

Math Boxes 2.7

---

1. Count up by 5s.

0,   5,   10,

_____ , _____ , _____ ,

_____ , _____ , _____

---

2. Make a tally for 16.

---

3. Record the time.

_____ o'clock

---

4. Circle the number that could be the mystery number.

8   9   10 11 12 13 14

9   11   13

Penny Grab Record Sheet

Round 1

I grabbed _____ pennies. | My partner grabbed _____ pennies.

I have _____ ¢. | My partner has _____ ¢.

Who has more? _____

Round 2

I grabbed _____ pennies. | My partner grabbed _____ pennies.

I have _____ ¢. | My partner has _____ ¢.

Who has more? _____

Round 3

I grabbed _____ pennies. | My partner grabbed _____ pennies.

I have _____ ¢. | My partner has _____ ¢.

Who has more? _____

Round 4

I grabbed _____ pennies. | My partner grabbed _____ pennies.

I have _____ ¢. | My partner has _____ ¢.

Who has more? _____

Use with Lesson 2.8.

Date

Math Boxes 2.8

1. Write the number that is 1 less.

_____ 12

_____ 16

_____ 21

2. Count back by 1s.

18,   17,   16,

_____, _____, _____,

_____, _____, _____,

_____, _____

3. Record the time.

_____ o'clock

4. Write the number.

_____   _____

Exploring Pennies and Nickels

Write the total amount. Then show the amount using fewer coins. Write ⓟ for penny and Ⓝ for nickel. (*Hint:* Exchange pennies for nickels.)

1.  _____¢

Show this amount using fewer coins.

2. ⓟⓟⓟⓟⓟⓟⓟⓟⓟ _____¢

Show this amount using fewer coins.

3.  _____¢

Show this amount using fewer coins.

Challenge

4. Ⓝⓟⓟⓟⓟⓟⓟ _____¢

Show this amount using fewer coins.

Use with Lesson 2.9.

Counting by 5s

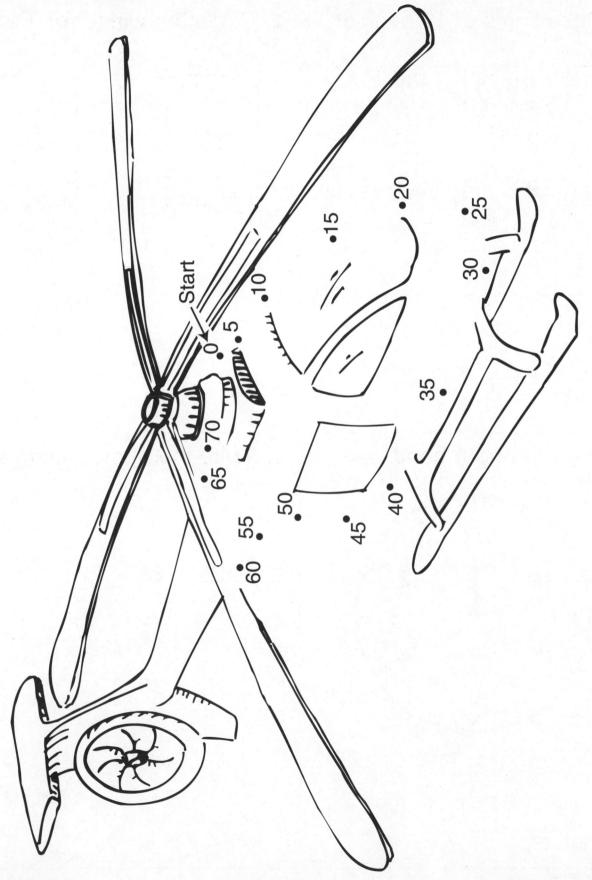

Math Boxes 2.9

1. Make sums of 10 pennies.

Left Hand	Right Hand
2	8
9	
	3

2. Circle the winner of *Top-It*.

18   17

3. Draw the hour hand.

9 o'clock

4. Write the number before.

_____ 19

_____ 23

_____ 31

_____ 36

Counting Pennies and Nickels

Write the total amount.

1.    _____ ¢

---

2.   _____ ¢

---

3.   _____ ¢

Challenge

Write the total amount.

4.

(P)	(N)	(P)		(P)	(P)
(P)			(P)	(N)	(P)
	(P)	(P)	(P)	(P)	

     _____ ¢

Show this amount using fewer coins.

5.

(N)	(N)	(P)		(P)
	(P)		(N)	
(P)		(N)		(P)

     _____ ¢

Show this amount using fewer coins.

Math Boxes 2.10

1. How many?

~~HHT~~ ~~HHT~~ ~~HHT~~ ||||

_____

2. Draw the hour hand.

3 o'clock

3. Record the total amount.

Ⓟ Ⓟ Ⓟ Ⓟ Ⓟ Ⓟ

_____¢

Use Ⓟ and Ⓝ to show this amount with fewer coins.

4. Count up by 2s.

0,   2,   4,

_____, _____, _____,

_____, _____, _____,

_____, _____

Nickel/Penny Grab Record Sheet

Round 1

We started with ____ Ⓝ and ____ Ⓟ for a total of ____ ¢.

I grabbed ____ Ⓝ and ____ Ⓟ.

My partner grabbed ____ Ⓝ and ____ Ⓟ.

I have ____ ¢. My partner has ____ ¢.

Who has more? _____

Round 2

We started with ____ Ⓝ and ____ Ⓟ for a total of ____ ¢.

I grabbed ____ Ⓝ and ____ Ⓟ.

My partner grabbed ____ Ⓝ and ____ Ⓟ.

I have ____ ¢. My partner has ____ ¢.

Who has more? _____

Round 3

We started with ____ Ⓝ and ____ Ⓟ for a total of ____ ¢.

I grabbed ____ Ⓝ and ____ Ⓟ.

My partner grabbed ____ Ⓝ and ____ Ⓟ.

I have ____ ¢. My partner has ____ ¢.

Who has more? _____

Math Boxes 2.11

1. Count the coins.

N N P

_____¢

2. Count up by 5s.

/0,   /5,   20,

_____, _____, _____,

_____, _____

3. Complete the table.

Before	Number	After
11	12	13
	14	
	16	
	18	

4. Draw the hour hand.

2 o'clock

Math Boxes 2.12

---

1. What time is it?

_____ o'clock

---

2. Make a tally for 23.

---

3. Fill in the blanks.

10,   20,   30,

_____, 50 , _____,

_____, 80 , _____,

100 , 110 , 120

---

4. Count the coins.

_____ ¢

School Store Mini-Poster 1

crayon
9¢

scissors
10¢

ball
25¢

gum
1¢

pencil
18¢

candy
5¢

eraser
7¢

Use with Lesson 2.13.

Counting Coins

1. Tell how much.

_____ ¢

_____ ¢

How much in all? _____ ¢

---

2. Buy 2 items from the School Store. Draw them below.

3. Under each item you drew, show how much it costs.
Use Ⓟs for pennies and Ⓝs for nickels.

4. Circle the item that costs more.

How much more does it cost? _____ ¢

Challenge

5. Draw 2 items that cost a total of 14¢.

Math Boxes 2.13

1. Count back by 1s.

21,   20,   19,

———— , ———— , ———— ,

———— , ———— , ———— ,

———— , ————

2. What time is it?

———— o'clock

3. Make sums of 10 pennies.

Left Hand	Right Hand
9	1
4	
	5

4. Use Ⓟ and Ⓝ to show the cost.

28¢

Math Boxes 2.14

1. Draw the hour hand.

6 o'clock

2. Record the total amount.

_____¢

Use Ⓝ to show this amount with fewer coins.

3. Complete the table.

Before	Number	After
7	8	9
	10	
	20	
	29	

4. Count up by 2s.

2,   4,   6,

_____ , _____ , _____ ,

_____ , _____ , _____ ,

_____ , _____

Patterns

1. Draw the next 2 shapes.
   Use your Pattern-Block Template.

2. Make up your own pattern.
   Then ask your partner to draw the next 2 shapes.

Challenge

3. Draw the next 3 shapes.

Use with Lesson 3.1.

1. Draw the hour hand.

4 o'clock

2. Count up by 2s.

___6___, ___8___, ___10___,

_____, _____, _____,

_____, _____, _____,

_____, _____

3. Draw dice dots for 6.

4. Record the total.

_____¢

Use ⓟ and Ⓝ to show this amount with fewer coins.

Odd and Even Patterns

Label **odd** or **even**. How many s?

Example

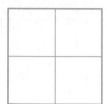

even

4

1. 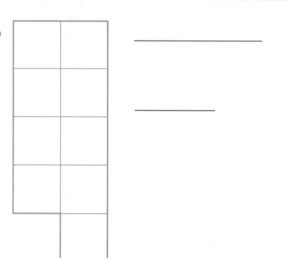 _____

_____

2. 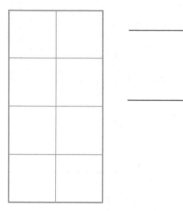 _____

_____

3. _____

_____

Label **odd** or **even**. How many ☆s?

4. ☆ ☆ ☆ ☆ ☆ ☆ ☆
☆ ☆ ☆ ☆ ☆ ☆ ☆   _____

_____

Challenge

5. ☆ ☆ ☆ ☆ ☆ ☆ ☆ ☆ ☆ ☆ ☆ ☆   _____
☆ ☆ ☆ ☆ ☆ ☆ ☆ ☆ ☆ ☆ ☆ ☆

_____

Use with Lesson 3.2.

Math Boxes 3.2

---

1. Make a tally for 25.

2. Write the number that is one more.

8 _____

14 _____

19 _____

28 _____

---

3. Make sums of 10 pennies.

Left Hand	Right Hand
3	7
4	
	8

4. Count up by 1s.

_36_ , _37_ , _____ ,

_____ , _____ , _____ ,

_____ , _____ , _____ ,

_____ , _____

The 3s Pattern

									0
1	2	3	4	5	6	7	8	9	10
11	12	13	14	15	16	17	18	19	20
21	22	23	24	25	26	27	28	29	30
31	32	33	34	35	36	37	38	39	40
41	42	43	44	45	46	47	48	49	50
51	52	53	54	55	56	57	58	59	60
61	62	63	64	65	66	67	68	69	70
71	72	73	74	75	76	77	78	79	80
81	82	83	84	85	86	87	88	89	90
91	92	93	94	95	96	97	98	99	100
101	102	103	104	105	106	107	108	109	110

Color the 3s pattern on the grid.
Fill in the missing numbers below.
Circle the ones digit in each 2-digit number.

___0___, ___3___, ___6___, _____, ___12___, _____,

_____, ___21___, _____, _____, ___30___, _____,

_____, _____, ___42___, _____, _____, ___51___

Date

1. Circle the winner of *Top-It.*

| 22 | 18 |

2. Write the total amount.

Ⓝ Ⓝ Ⓟ Ⓟ

_____¢

3. What time is it?

_____ o'clock

4. Count up by 5s.

__/5__ , __20__ , __25__ ,

_____ , _____ , _____ ,

_____ , _____ , _____

Math Boxes 3.4

1. Count back by 1s.

_43_ , _42_ , _41_ ,

_____ , _____ , _____ ,

_____ , _____ , _____ ,

_____ , _____

2. What time is it?

_____ o'clock

3. Draw what comes next.

△ ◯ △ ◯ ___ ___

✕ | | ✕ ___ ___

4. Write the number that is one less.

_____ 16   _____ 20

_____ 24   _____ 39

Number-Line Skip Counting

1. Show counts by 2s.

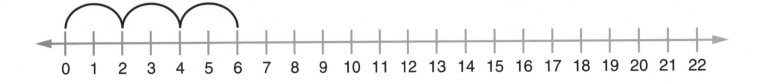

2. Show counts by 5s.

3. Show counts by 10s.

4. Show counts by 3s.

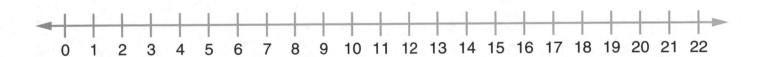

Date

Number-Line Hops

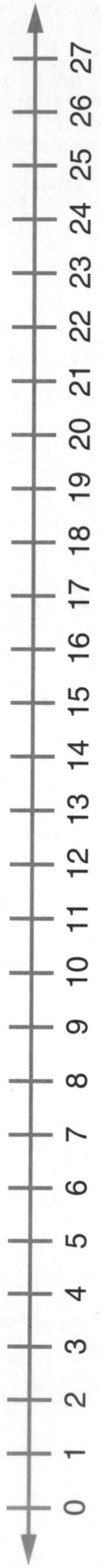
0
1
2
3
4
5
6
7
8
9
10
11
12
13
14
15
16
17
18
19
20
21
22
23
24
25
26
27

1. Start at 0. Count up 6 hops. Then count up 5 more hops.

Where do you end up? ______

2. Start at 9. Count up 2 hops.

Where do you end up? ______

3. Start at 14. Count up 0 hops.

Where do you end up? ______

4. Start at 0. Count up 5 hops. Then count back 3 hops.

Where do you end up? ______

5. Start at 18. Count back 9 hops.

Where do you end up? ______

6. Start at 11. Count up to 17.

How many hops is it from 11 to 17? ______

Use with Lesson 3.5.

1. Count up by 2s.

_12_ , _14_ , _16_ ,

_____ , _____ , _____ ,

_____ , _____ , _____

_____ , _____

2. Write the total amount.

Ⓝ Ⓝ Ⓝ Ⓟ

_____ ¢

3. Draw the hour hand.

9 o'clock

4. Match the numbers to the number words.

1 - - - - - - - five

2 - - - - - - one

3   four

4   two

5   three

Adding and Subtracting on the Number Line

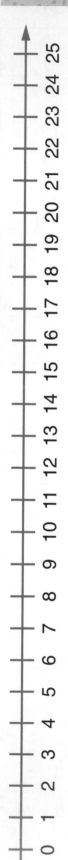

1. Start at 6. Count up 2 hops. Where do you end up? ____

$6 + 2 = $ ____

2. Start at 4. Count up 9 hops. Where do you end up? ____

$4 + 9 = $ ____

3. Start at 15. Count back 7 hops. Where do you end up? ____

$15 - 7 = $ ____

4. Start at 18. Count back 8 hops. Where do you end up? ____

$18 - 8 = $ ____

Challenge

5. $5 + 8 = $ ____

6. $11 - 8 = $ ____

7. $3 + 13 = $ ____

Use with Lesson 3.6.

Math Boxes 3.6

1. Draw the hour hand.

5 o'clock

2. Odd or even?

_____ _____

_____ _____

3. Complete the table.

Before	Number	After
24	25	26
	29	
	33	
	37	
	40	

4. Count up by 10s.

_20_ , _30_ , _40_ ,

_____ , _____ , _____ ,

_____ , _____ , _____ ,

_____ , _____

Telling Time

1. Record the time.

half-past _____ o'clock

half-past _____ o'clock

_____ o'clock

_____ o'clock

2. Draw the hour hand and the minute hand.

half-past 1 o'clock

7 o'clock

Use with Lesson 3.7.

Number Lines

Complete the following number lines.

1.

_17_   ___   ___   ___   _21_   ___   ___

2.

___   _52_   ___   ___   ___   ___   _57_

3.

___   ___   ___   _69_   ___   _71_

4.

___   _98_   ___   ___   ___   _102_   ___

Make up your own.

5.

___   ___   ___   ___   ___   ___

Challenge

6.

___   _-1_   _0_   _1_   ___   ___   ___

Math Boxes 3.7

1. Make sums of 10 pennies.

Left Hand	Right Hand
2	8
5	
	9

2. Make a tally for 27.

Odd or even? _____

3. Draw what comes next.

_____ _____

_____ _____

4. Write the total amount.

_____ ¢

Frames and Arrows

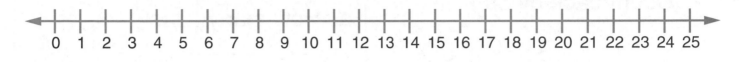

0 1 2 3 4 5 6 7 8 9 10 11 12 13 14 15 16 17 18 19 20 21 22 23 24 25

1.

Rule
Count up by 2s

| 10 | 12 | | | | |

2.

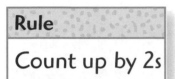

Rule
Add 5

| 0 | 5 | 10 | | | |

3.

Rule
Count back by 2s

| 13 | 11 | | | | 3 |

4.

Rule
Subtract 3

| 15 | | | | | |

Math Boxes 3.8

1. Count back by 5s.

_____40_____, _____35_____, _____30_____,

_____, _____, _____,

_____, _____, _____,

_____

2. How many 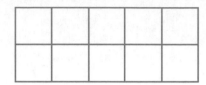s?

_____

Odd or even? _____

3. How much money?

_____¢

4. Use your number line.
Start at 0. Count up 6 hops.

You end at _____.

$0 + 6 =$ _____

More Frames and Arrows

1. Fill in the frames.

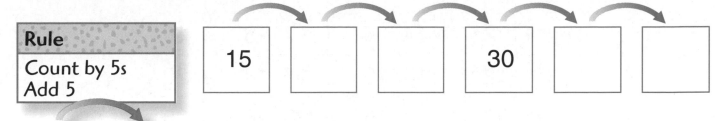

Rule
Count by 5s Add 5

15 | | | 30 | |

2. Fill in the rule.

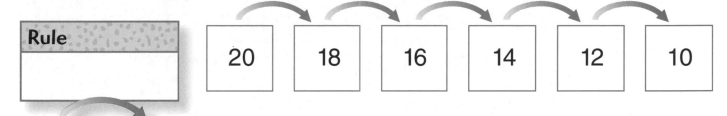

Rule

20 | 18 | 16 | 14 | 12 | 10

3. Fill in the rule and the frames.

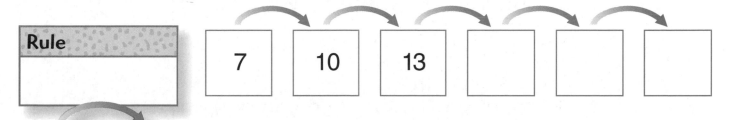

Rule

7 | 10 | 13 | | |

4. Fill in the rule and the frames.

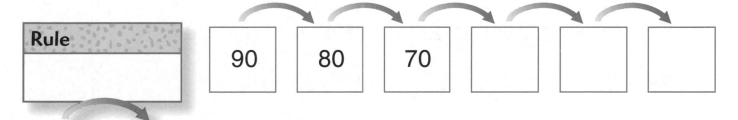

Rule

90 | 80 | 70 | | |

5. Make up your own.

Rule

| | | | | |

Adding on the Number Grid

									0
1	2	3	4	5	6	7	8	9	10
11	12	13	14	15	16	17	18	19	20
21	22	23	24	25	26	27	28	29	30
31	32	33	34	35	36	37	38	39	40
41	42	43	44	45	46	47	48	49	50
51	52	53	54	55	56	57	58	59	60
61	62	63	64	65	66	67	68	69	70

1. Start at 25. Count up 3. Where do you end up?   _____

$25 + 3 =$ _____

2. Start at 19. Count up 6. Where do you end up?   _____

$19 + 6 =$ _____

3. Start at 38. Count up 2. Where do you end up?   _____

$38 + 2 =$ _____

4. Start at 57. Count up 10. Where do you end up?   _____

$57 + 10 =$ _____

5. $29 + 20 =$ _____   **6.** $25 + 15 =$ _____

50 (fifty)

1. Count back by 10s.

_90_ , _80_ , _70_ ,

_____ , _____ , _____ ,

_____ , _____ , _____

2. Record the time.

half-past _____ o'clock

3.

Rule
Count by 5s

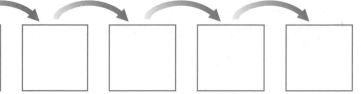

15

4. Complete the number line.

_16_   _17_   ___   ___   ___   ___   ___   ___

Calculator Counts

									0
(1)	2	3	4	5	6	7	8	9	10
11	12	13	14	15	16	17	18	19	20
21	22	23	24	25	26	27	28	29	30
31	32	33	34	35	36	37	38	39	40
41	42	43	44	45	46	47	48	✗49	50

Do the following counts on a calculator:

1.   Start at: 1
Count: up
By: 2s

Circle these numbers on the grid.

2.   Start at: 49
Count: back
By: 3s

Mark these numbers on the grid with an X.

Challenge

3. List the numbers that have a circle and an X.

_____, _____, _____, _____, _____, _____, _____, _____, _____

These are counts by _____.

Subtracting on the Number Grid

									0
1	2	3	4	5	6	7	8	9	10
11	12	13	14	15	16	17	18	19	20
21	22	23	24	25	26	27	28	29	30
31	32	33	34	35	36	37	38	39	40
41	42	43	44	45	46	47	48	49	50
51	52	53	54	55	56	57	58	59	60
61	62	63	64	65	66	67	68	69	70

1. Start at 35. Count back 2. Where do you end up? _____

$35 - 2 =$ _____

2. Start at 27. Count back 5. Where do you end up? _____

$27 - 5 =$ _____

3. Start at 48. Count back 10. Where do you end up? _____

$48 - 10 =$ _____

4. Start at 65. Count back 15. Where do you end up? _____

$65 - 15 =$ _____

5. $52 - 20 =$ _____   **6.** $46 - 16 =$ _____

Math Boxes 3.10

---

1. Fill in the rule and the missing numbers.

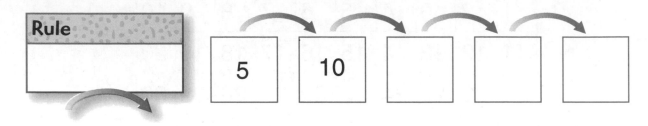

Rule

| 5 | 10 | | | |

---

2. Count up by 10s.

___30___ , ___40___ , ___50___ , _____ , _____ ,

_____ , _____ , _____ , _____

---

3. Odd or even?

_____   _____

---

4. How much money?

_____ ¢

Use Ⓟ and Ⓝ to show this amount with fewer coins.

Coin Exchange

Show each amount using fewer coins.
Write Ⓟ for penny, Ⓝ for nickel, and Ⓓ for dime.

1.

2.

3.

4.

Frames and Arrows

Fill in the missing numbers or the missing rule.
Use your calculator or a number grid to help you.

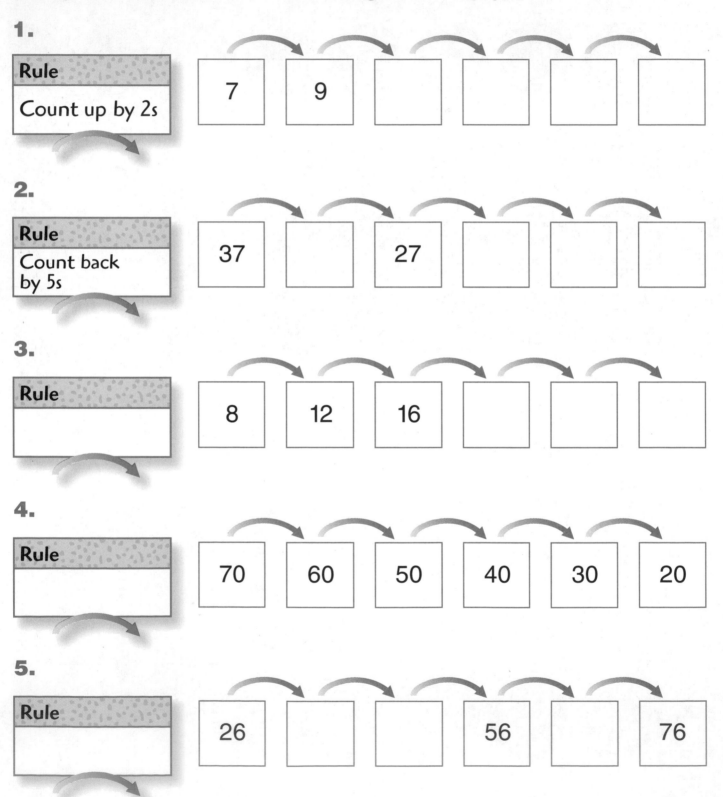

1.

Rule

Count up by 2s

| 7 | 9 | | | | |

2.

Rule

Count back
by 5s

| 37 | | 27 | | | |

3.

Rule

| 8 | 12 | 16 | | | |

4.

Rule

| 70 | 60 | 50 | 40 | 30 | 20 |

5.

Rule

| 26 | | | 56 | | 76 |

Use with Lesson 3.11.

Math Boxes 3.11

1. Circle the even numbers.

3   6   9   12

15   18   21   24

2. Complete the number line.

<u>44</u>   <u>45</u>   ___   ___   ___   ___   ___   ___

3. Use your number line.
Start at 3. Count up 5 hops.

You end at _____.

3 + 5 = _____

4. What time is it?

half-past _____ o'clock

How Much Money?

How much money?
Use your coins.

Example

P	N	D
1¢	5¢	10¢
$0.01	$0.05	$0.10
a penny	a nickel	a dime

___35___ ¢ or $ _0.35_

How much money? Use your coins.

1.  $_____

---

2. $_____

3.

   _____¢

---

4. D D N N N P P $_____

    Use with Lesson 3.12.

Date

1. Write the number for:

~~HHT~~ ~~HHT~~ ~~HHT~~ ~~HHT~~ ~~HHT~~ ~~HHT~~
~~HHT~~ ~~HHT~~ //

_____

Odd or even? _____

2. Use your number grid.
Start at 52. Count back 8.

You end at _____.

$52 - 8 =$ _____

3. Complete the number line.

___ /0 // ___ ___ ___ ___

4.

Rule		20		12	
−4					

Calendar Patterns

Fill out the calendar for this month.

Month _____

Sunday	Monday	Tuesday	Wednesday	Thursday	Friday	Saturday

Circle all the even numbers.
Make a ✓ next to each odd number.
Look for patterns in the even and odd numbers.
Discuss the patterns with your partner.

Use with Lesson 3.13.

Math Boxes 3.13

1. Write the total amount.

_____¢

2.

Rule
Count by 3s

3	6	9		

3. Draw the hour hand and the minute hand.

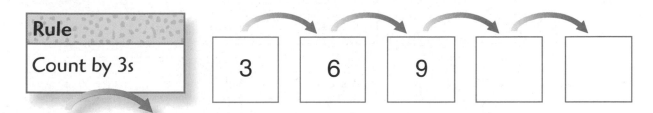

half-past 12 o'clock

4. Use a calculator. Count by 10s.

___80___, ___90___, _____,

_____, _____, _____,

_____, _____

Domino Parts and Totals

Write 3 numbers for each domino.

Example

Part	Part
4	2
Total	
6	

1.

Part	Part
Total	

2.

Part	Part
Total	

3.

Part	Part
Total	

4.

Part	Part
Total	

5.

Part	Part
Total	

6. Draw dots in the domino. Write 3 numbers in the diagram.

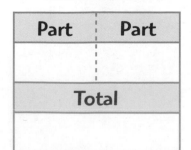

Part	Part
Total	

Challenge

7. Find the missing part. Draw dots in the domino.

Part	Part
3	
Total	
8	

Use with Lesson 3.14.

1. Make a tally for 28.

Odd or even? _____

2. Draw the hour hand and the minute hand.

half-past 2 o'clock

3. Use your number grid.

Start at 45. Count up 13.

You end at _____.

45 + 13 = _____

4. Write the total amount.

Ⓓ Ⓓ Ⓓ Ⓝ Ⓝ Ⓝ Ⓟ Ⓟ Ⓟ

_____ ¢ or $_____

Use Ⓟ, Ⓝ and Ⓓ to show this amount another way.

Math Boxes 3.15

1. Draw the hour hand and the minute hand.

half-past 5 o'clock

2. Make sums of 10 pennies.

Left Hand	Right Hand
8	2
6	
	3

3.

Rule
Count by 2s

4. Complete the number line.

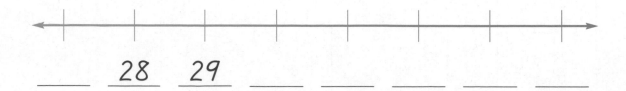

28   29

Use with Lesson 3.15.

Reading Thermometers

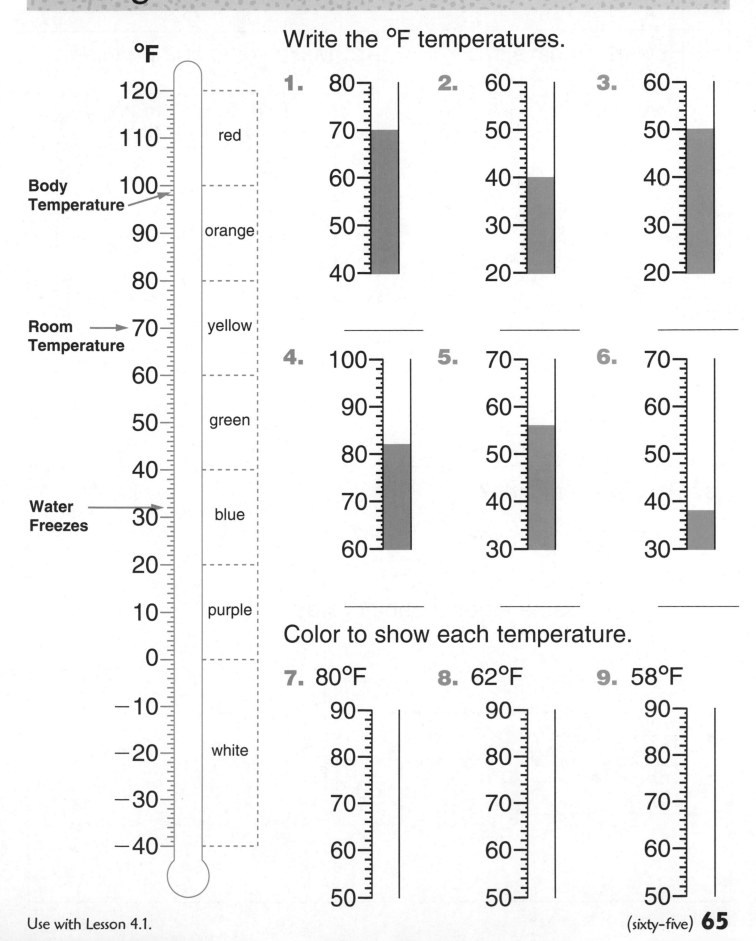

°F

120
110 — red
Body Temperature → 100
90 — orange
80
Room Temperature → 70 — yellow
60
50 — green
40
Water Freezes → 30 — blue
20
10 — purple
0
−10
−20 — white
−30
−40

Write the °F temperatures.

1. [thermometer 40–80, reading 70]

2. [thermometer 20–60, reading 40]

3. [thermometer 20–60, reading 50]

4. [thermometer 60–100, reading 82]

5. [thermometer 30–70, reading 56]

6. [thermometer 30–70, reading 38]

Color to show each temperature.

7. 80°F [thermometer 50–90]

8. 62°F [thermometer 50–90]

9. 58°F [thermometer 50–90]

Math Boxes 4.1

1. How many days in

2 weeks? _____

2. Make sums of 9 pennies.

Left Hand	Right Hand
4	5

3. How much money?

Ⓓ Ⓓ Ⓓ Ⓝ Ⓟ Ⓟ Ⓟ _____¢

Show the same amount another way.

4.

Rule
Count back by 10s

70 → 60 → ☐ → ☐ → ☐

Use with Lesson 4.1.

My Body and Units of Measure

Measure some objects. Record your measurements.

Unit	Picture	Object	Measurements
digit			about ____ digits
yard			about ____ yards
hand			about ____ hands
pace			about ____ paces
cubit			about ____ cubits
arm span (or fathom)			about ____ arm spans
foot			about ____ feet
hand span			about ____ hand spans

My Height

Things that are taller than I am.

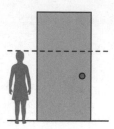

Things that are about the same size as I am.

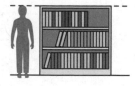

Things that are shorter than I am.

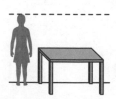

Use with Lesson 4.2.

Math Boxes 4.2

1. Use your number grid.

Start at 48. Count up 15.

You end at _____.

48 + 15 = _____

2. Draw the hands.

half-past 8 o'clock

3. Circle the winner of *Domino Top-It.*

4. Most popular pet _____

How many like snakes?

_____

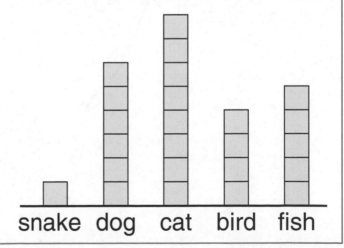

Favorite Pets, Mrs. Lee's Class

snake dog cat bird fish

Two-Fisted Penny Addition Summary

5	
Left	Right
2	3
3	2
1	4
4	1
0	5
5	0

7	
Left	Right

8	
Left	Right

9	
Left	Right

10	
Left	Right

6	
Left	Right

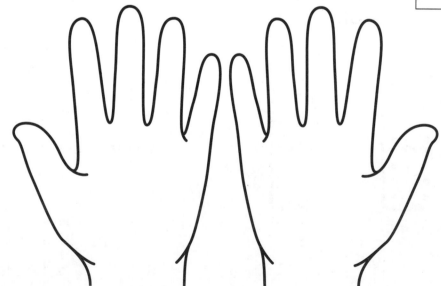

Two-Fisted Penny Addition Summary (cont.)

11		12		13		14		15	
Left	Right	Left	Right	Left	Right	Left	Right	Left	Right

Two-Fisted Penny Addition Summary (cont.)

16		17		18		19		20	
Left	Right	Left	Right	Left	Right	Left	Right	Left	Right

Use with Lesson 4.2.

My Foot and the Standard Foot

Measure two objects with the cutout of your foot.
Draw pictures of the objects, or write their names.

1. I measured

It is about _____ _____ feet.

(your name)

2. I measured

It is about _____ _____ feet.

(your name)

Measure two objects with the foot-long foot.
Sometimes it is called the *standard foot*.

3. I measured

It is about _____ feet.

4. I measured

It is about _____ feet.

Math Boxes 4.3

1. Complete the table.

Before	Number	After
27	28	29
	35	
	40	
	101	

2. Show 26¢.

Use Ⓓ, Ⓝ, and Ⓟ.

3. Count back by 2s.

_36_ , _34_ , _32_ ,

_____ , _____ , _____ ,

_____ , _____ , _____ ,

_____ , _____

4. What is the temperature today?

_____ °F

Odd or even?

_____

Inches

Pick 4 short objects to measure. Draw or name them.
Then measure them.

1.

About _____ inches long

2.

About _____ inches long

3.

About _____ inches long

4.

About _____ inches long

Math Boxes 4.4

---

1. Complete the number line.

___ _36_ _37_ ___ ___ ___

---

2. Make a tally for 24.

Odd or even? _____

---

3. Draw the hands.

half-past 2 o'clock

---

4. There are 13 pennies.

You grab 4.

Your partner must

have _____.

Measuring in Inches

Choose two objects to measure. Estimate each object's length. Measure the objects to the nearest inch.

Object (Name it or draw it)	My Estimate	My Measurement
	about _____ inches	about _____ inches
	about _____ inches	about _____ inches

Measure each line segment.

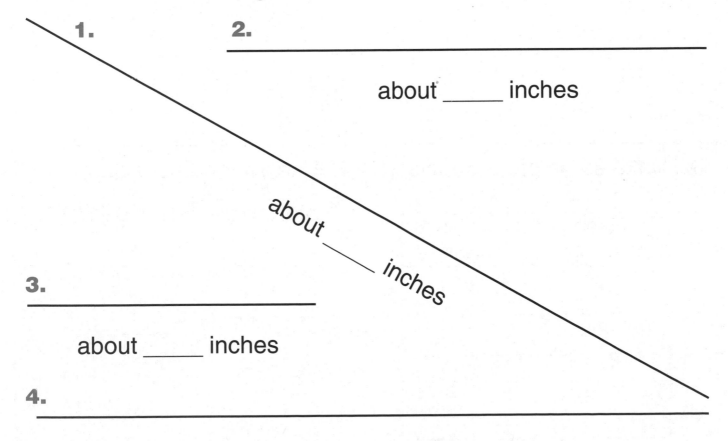

1.

2. _____

about _____ inches

about _____ inches

3. _____

about _____ inches

4. _____

about _____ inches

Draw a line segment about

5. 3 inches long

6. 5 inches long

Use with Lesson 4.5.

Math Boxes 4.5

1. Fill in the rule and the missing numbers.

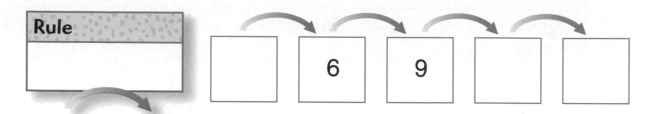

Rule

| | 6 | 9 | | |

2. Write the number that is one more.

33 _____   71 _____

46 _____   109 _____

3. Make sums of 15 pennies.

Left Hand	Right Hand
10	5
8	
	6

4. Use your number grid.

Start at 36. Count back 14.

You end at _____.

36 − 14 = _____

Date

Measuring Parts of the Body

Record your wrist size below.

1. Wrist 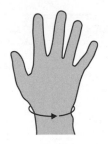 It is about _____ inches.

Measure these other parts of your body. Work with a partner.

2. Elbow 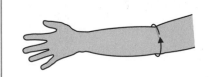 It is about _____ inches.

3. Neck 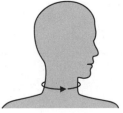 It is about _____ inches.

4. Head 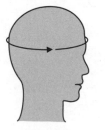 It is about _____ inches.

5. Hand span 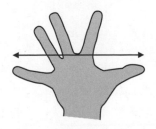 It is about _____ inches.

Historical Units of Measure

Measure these parts of your body. Work with a partner.

Historical Unit	Picture	Me
digit (finger width)		about _____ inches
hand		about _____ inches
cubit		about _____ inches
yard		about _____ inches
fathom (arm span)		about _____ inches
pace (heel to toe)		about _____ inches

Use with Lesson 4.6.

Domino Dots

Draw the missing dots on each domino.
Then write the total number of dots.

1. _____

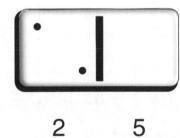

3   2

2. _____

4   3

3. _____

5   4

4. _____

2   5

5. _____

2   4

6. _____

3   5

Challenge

7.   *10*

_____   _____

8.   *8*

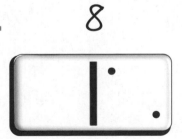

_____   _____

Make up your own.

9. _____

_____   _____

Math Boxes 4.6

1. Use a number grid. Count by 10s.

_8_ , _18_ , ———— ,

———— , ———— , ———— ,

———— , ———— , ———— ,

———— , ————

2. Record the time.

half-past ———— o'clock

3. Tim has 10¢. Jan has 5¢.

Who has more?

————

How much more?

————

4. Measure your calculator.

How long is it?

It is about ———— inches long.

Use with Lesson 4.6.

Measuring Height

1. Today's date is _____.

   My height is _____ inches.

2. This is a bar graph. It shows the heights of children in my class.

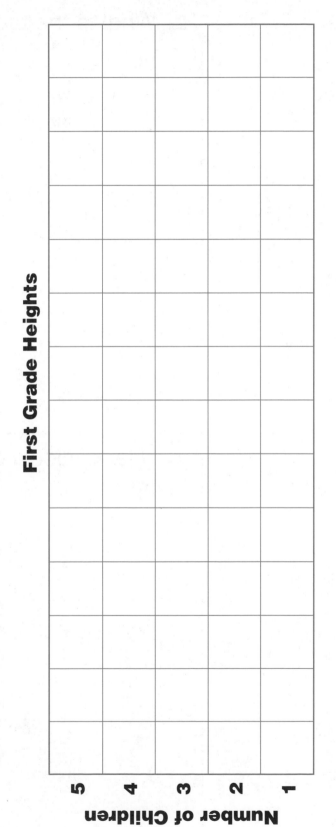

First Grade Heights

Inches Tall

Number of Children
5
4
3
2
1

3. This is the "typical" height for first graders in my class:

   About _____ inches.

Math Boxes 4.7

1. Draw the hands.

half-past 3 o'clock

2. What is the temperature

today? _____ °F

Odd or even?

_____ °F

3. Show 47¢.

Use Ⓓ, Ⓝ, and Ⓟ.

4. Use a calculator. Count up
by 7s.

___0___, ___7___, __14__,

_____, _____, _____,

_____, _____, _____,

_____, _____

Telling Time

Record the time.

1.

_____ o'clock

2.

half-past _____ o'clock

3.

quarter-past _____ o'clock

4.

quarter-to _____ o'clock

Challenge

Draw the hour hand and the minute hand.

5.

half-past 3 o'clock

6.

quarter-to 5 o'clock

Math Boxes 4.8

1. Write some odd numbers in this box.

2. Fill in the rule and the missing numbers.

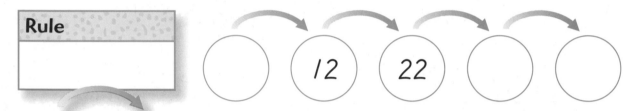

Rule

12   22

3. For each domino:

Draw the missing dots.

Find the total number of dots.

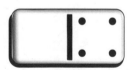

5 + 4 = _____

6 + 3 = _____

4. Measure to the nearest inch.

_____

It is about _____ inches long.

_____

It is about _____ inches long.

Use with Lesson 4.8.

School-Year Timeline

Think about these times of the year.
Draw pictures of things that happen during each of these months.

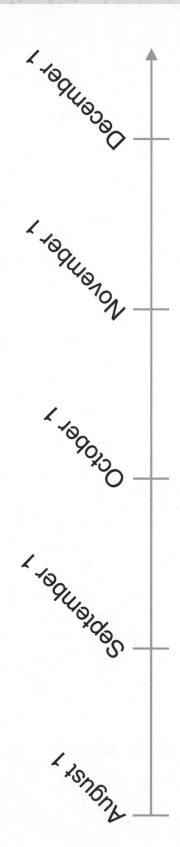

August 1

September 1

October 1

November 1

December 1

Domino Parts and Totals

Find the totals.

Example

12

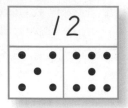

1.

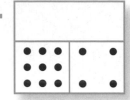

2.

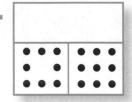

3.

4.	Total	
	Part	**Part**
	7	4

5.	Total	
	Part	**Part**
	2	9

6.	Total	
	Part	**Part**
	6	3

7.	Total	
	Part	**Part**
	8	8

8.	Total	
	Part	**Part**
	7	7

9.	Total	
	Part	**Part**
	8	3

Challenge

Find the missing part.

10.	Total	
	12	
	Part	**Part**
	6	

11.	Total	
	14	
	Part	**Part**
		5

12.	Total	
	15	
	Part	**Part**
		7

Use with Lesson 4.9.

Math Boxes 4.9

1. Use your number grid.

Start at 59. Count back 20.

You end at _____.

_____ = 59 − 20

2. How much money?

Ⓓ Ⓓ Ⓓ Ⓓ Ⓓ Ⓝ Ⓟ Ⓟ Ⓟ

_____ ¢ or $_____

Show this amount another way.

3. How old were you three years ago?

_____

Odd or even?

_____

4. Measure your journal.

How long is it?

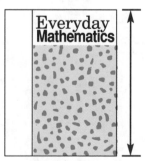

It is about _____ inches long.

Math Boxes 4.10

1. Circle the winner of *Domino Top-It*.  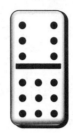	**2.** Make a tally for 37. Odd or even? _____
3. Draw the hands.  half-past 10 o'clock	**4.** Use your calculator. Count up by 3s. ___0___ , ___3___ , ___6___ , _____ , _____ , _____ , _____ , _____ , _____ , _____ , _____

Domino Sums

Find the sums. Use the dominoes below to help you.

1. 2 + 4 = ____

2. 3 + 1 = ____

3. ____ = 1 + 5

4. 2 + 2 = ____

5. ____ = 3 + 3

6. ____ = 4 + 1

7. 4
     + 2

8. 5
     + 1

9. 1
     + 3

Challenge

10. Draw a domino of your choice. Write a fact to go with it.

____ + ____ = ____

Math Boxes 4.11

1. Find the missing parts.

Total
8

Part	Part
3	

Total
14

Part	Part
	6

2. Make sums of 16 pennies.

Left Hand	Right Hand
6	10
8	
	9

3. How many school days so far?

_____

In 10 more days it will be day

_____

4. Show 32¢ with the fewest coins.

Use Ⓓ, Ⓝ, and Ⓟ.

Color by Number

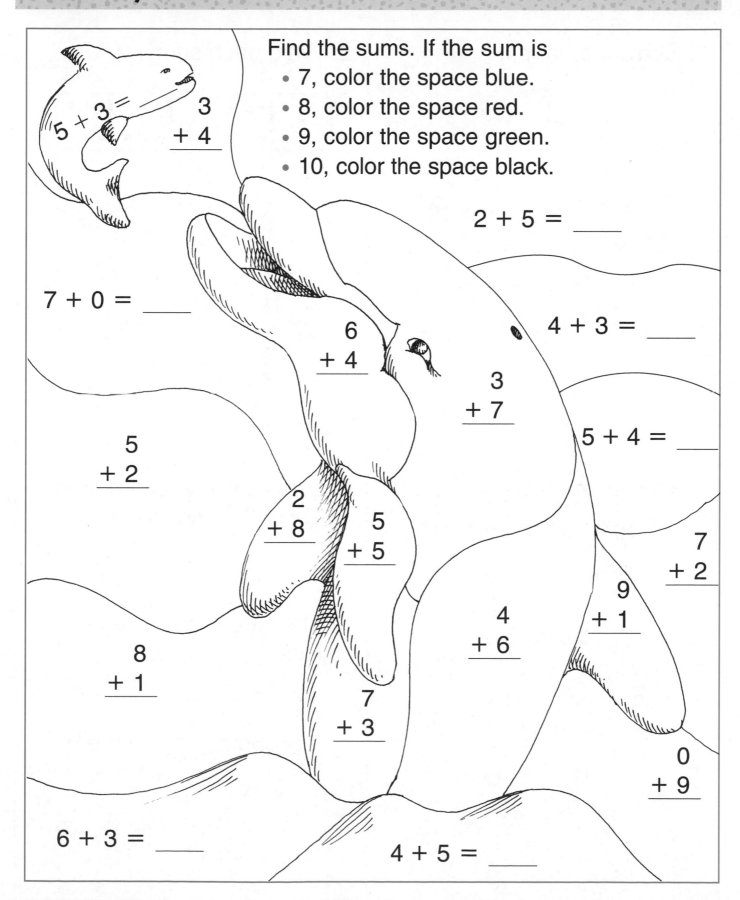

Find the sums. If the sum is
- 7, color the space blue.
- 8, color the space red.
- 9, color the space green.
- 10, color the space black.

$5 + 3 =$

$\begin{array}{r} 3 \\ + 4 \\ \hline \end{array}$

$2 + 5 =$ ___

$7 + 0 =$ ___

$\begin{array}{r} 6 \\ + 4 \\ \hline \end{array}$

$4 + 3 =$ ___

$\begin{array}{r} 3 \\ + 7 \\ \hline \end{array}$

$5 + 4 =$ ___

$\begin{array}{r} 5 \\ + 2 \\ \hline \end{array}$

$\begin{array}{r} 2 \\ + 8 \\ \hline \end{array}$

$\begin{array}{r} 5 \\ + 5 \\ \hline \end{array}$

$\begin{array}{r} 7 \\ + 2 \\ \hline \end{array}$

$\begin{array}{r} 9 \\ + 1 \\ \hline \end{array}$

$\begin{array}{r} 8 \\ + 1 \\ \hline \end{array}$

$\begin{array}{r} 4 \\ + 6 \\ \hline \end{array}$

$\begin{array}{r} 7 \\ + 3 \\ \hline \end{array}$

$\begin{array}{r} 0 \\ + 9 \\ \hline \end{array}$

$6 + 3 =$ ___

$4 + 5 =$ ___

1. Draw the hands.

quarter-to 5 o'clock

2. Write each sum.

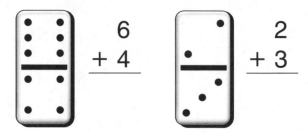

1 + 5 = _____   6 + 6 = _____

$$\begin{array}{r} 6 \\ + 4 \\ \hline \end{array}$$ $$\begin{array}{r} 2 \\ + 3 \\ \hline \end{array}$$

3. How much money?

Ⓓ Ⓝ Ⓝ Ⓝ Ⓝ Ⓝ Ⓟ Ⓟ
Ⓟ Ⓟ Ⓟ Ⓟ

_____¢ or $_____

Use Ⓟ, Ⓝ, and Ⓓ to show this amount another way.

4. Draw a line segment that is about 2 inches long.

Measure to the nearest inch.

_____

It is about _____ inch(es) long.

Math Boxes 4.13

1. Record the time.

quarter-after _____ o'clock

2. Use your number grid.

Start at 71. Count up 19.

You end at _____.

71 + 19 = ___

3. Find the sums.

6 + 1 = ___   _____ = 1 + 8

_____ = 2 + 2   _____ = 0 + 4

Circle the odd sums.

4.

Rule
Count back by 10s

| 44 | 34 | 24 | | |

Tens-and-Ones Mat

Tens 10s	Ones 1s

 Use with Lesson 5.1.

Tens-and-Ones Riddles

Solve the riddles. Use your base-10 blocks to help you.
Example

3☐ and 2▯. What am I? _23_

1. 5☐ and 6▯. What am I? ____

2. 2▯ and 7☐. What am I? ____

3. 4 cubes and 6 longs. What am I? ____

4. 7 longs and 0 cubes. What am I? ____

Here are some harder riddles.
You will need to trade to find the answers.

5. 1 long and 11 cubes. What am I? ____

6. 14 cubes and 2 longs. What am I? ____

7. Make up your own riddle.
   Ask a friend to solve your riddle.

Frames and Arrows

Fill in the missing numbers or the missing rule.
Use your calculator or a number grid to help you.

1.

Rule

Add 2

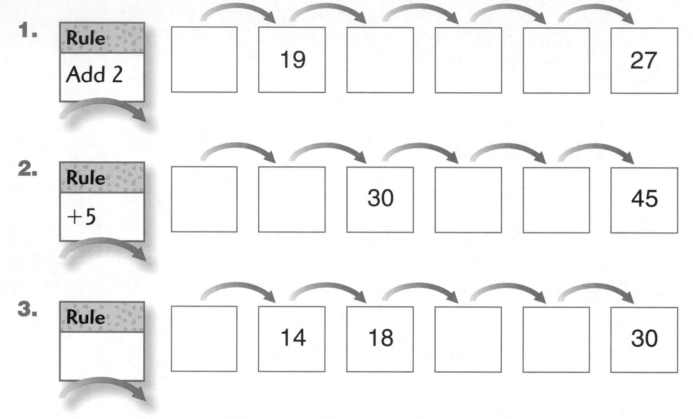

| | 19 | | | | 27 |

2.

Rule

+5

| | | 30 | | | 45 |

3.

Rule

| | 14 | 18 | | | 30 |

4.

Rule

| | | 42 | 32 | | 12 |

Challenge

5.

Rule

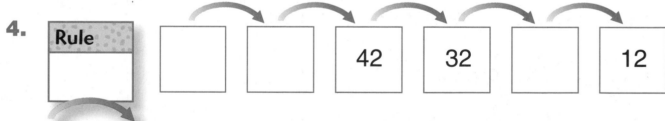

| | | 38 | | 34 | |

Use with Lesson 5.1.

Math Boxes 5.1

1. Write one more.

36 _____

45 _____

61 _____

83 _____

2. Make sums of 12 pennies.

Left Hand	Right Hand
7	
	6
	3

3. Draw what comes next.

4. Find the sums.

5 + 4 = _____   _____ = 3 + 7

Place-Value Mat

	Hundreds
	Tens
	Ones

Math Boxes 5.2

1. What time is it?

quarter-before _____ o'clock

2. Start at 57. Count up 30.

You end at _____.

$$\begin{array}{r} 57 \\ + \ 30 \\ \hline \end{array}$$

3. How much money?

Ⓓ Ⓓ Ⓝ Ⓝ Ⓝ Ⓝ Ⓝ Ⓟ

_____ ¢ or $_____

4. Circle the winner of *Top-It*.

| 20 | 22 |

Tens–and–Ones Trading Game

Materials
- ❑ a die
- ❑ base-10 blocks (longs and cubes)
- ❑ a Tens-and-Ones Mat (*Math Journal 1,* page 96) for each player

Players   2 or 3

Directions

Take turns rolling the die.

If you roll a 1, take 1 long.

If you roll a 2, take 2 longs.

If you roll a 3, take 3 cubes.

If you roll a 4, take 4 cubes.

If you roll a 5, take 5 cubes.

If you roll a 6, take 6 cubes.

Put your blocks on your Tens-and-Ones Mat.
If you can, trade 10 cubes for 1 long.

The winner is the first player to get 10 longs.

Another Way to Play

Place 10 tens on your Tens-and-Ones Mat.
Take turns rolling the die and taking pieces off the mat.
You must get the exact number to take the last pieces off.
The winner is the first to take all the pieces off.

Use with Lesson 5.3.

Math Boxes 5.3

1. Write the sums.

$2 + 4 =$ _____

_____ $= 3 + 3$

_____ $= 1 + 5$

_____ $= 4 + 2$

2. Measure to the nearest inch.

_____

_____ inches

_____

_____ inches

3.

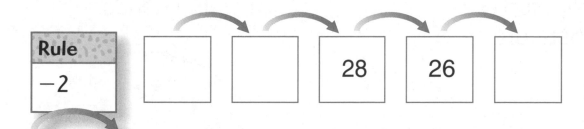

Rule
−2

| | | 28 | 26 | |

4. Solve the riddles.

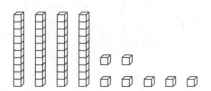

What am I? _____

What am I? _____

Math Boxes 5.4

---

1. Count up by 10s. You may use a calculator.

77,   87,   _____,

_____, _____, _____,

_____, _____, _____,

_____, _____

2.

$$7 + 3$$ $$2 + 6$$

$$2 + 2$$ $$8 + 1$$

---

3. Record today's temperature.

_____ °F

Odd or even?

_____

4. Draw the hands.

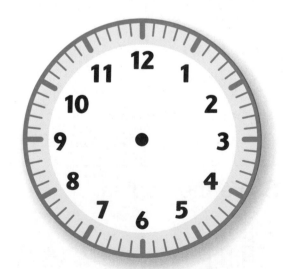

quarter-to 6 o'clock

Animal Weights

Find the total weight of the 2 animals.

"lb" means "pound."

1.

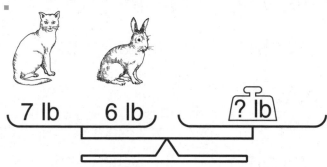

7 lb   6 lb   ? lb

7 lb + 6 lb = _____ lb

2.

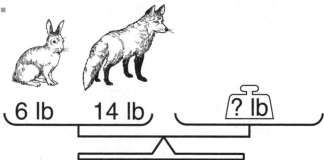

6 lb   14 lb   ? lb

6 lb + 14 lb = _____ lb

3.

7 lb   23 lb   ? lb

7 lb + 23 lb = _____ lb

4.

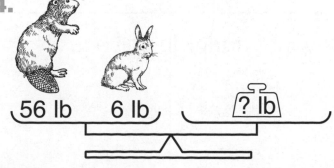

56 lb   6 lb   ? lb

56 lb + 6 lb = _____ lb

More Animal Weights

Take 2 animal cards. Write each animal's name and weight. Find their total weight.

1. The _____ weighs _____ pounds.

The _____ weighs _____ pounds.

Together, they weigh _____ pounds.

2. The _____ weighs _____ pounds.

The _____ weighs _____ pounds.

Together, they weigh _____ pounds.

3. The _____ weighs _____ pounds.

The _____ weighs _____ pounds.

Together, they weigh _____ pounds.

4. The _____ weighs _____ pounds.

The _____ weighs _____ pounds.

Together, they weigh _____ pounds.

5. Make up an animal story. Ask your partner to solve it.

Use with Lesson 5.5.

Math Boxes 5.5

---

1. Show 43¢ in two ways.
Use Ⓓ, Ⓝ, and Ⓟ.

---

2. Use your number grid.
Start at 33. Count back 9.

You end at _____.

33
− 9
_____

---

3. Write the sums.

_____ = 2 + 5

5 + 7 = _____

---

4. Complete the table.

Before	Number	After
10	11	12
	17	
	59	
	85	
	100	

"Less Than" and "More Than" Number Models

Write < for "is less than" and > for "is more than."

1. 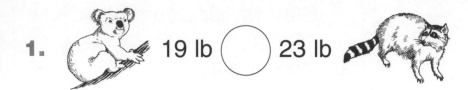 19 lb ◯ 23 lb

2. 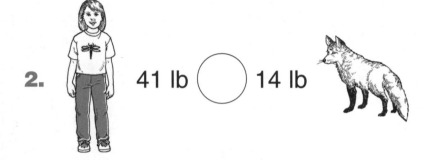 41 lb ◯ 14 lb

3.  75 lb ◯ 56 lb

Challenge

4. 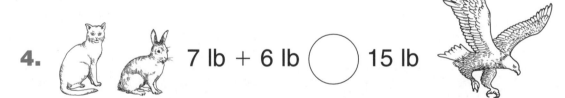 7 lb + 6 lb ◯ 15 lb

5. 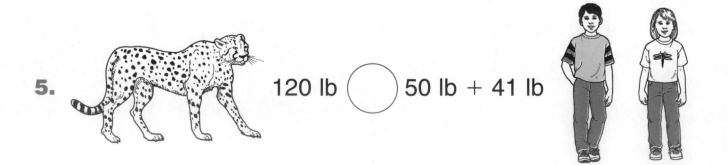 120 lb ◯ 50 lb + 41 lb

Use with Lesson 5.6.

Ordering Animals by Weight

Order your animal cards from largest to smallest weight.
Record the results below.

1st	*cheetah*	*120*	lb
2nd			lb
3rd			lb
4th			lb
5th	*8-year-old boy*	*50*	lb
6th			lb
7th			lb
8th			lb
9th			lb
10th	*fox*	*14*	lb
11th			lb
12th			lb

Math Boxes 5.6

1. Write <, >, or =.

3 ☐ 13

17 ☐ 15

24 ☐ 42

28 ☐ 26

2. Draw the hands.

half-past 10 o'clock

3. How old will you be in 15 years?

_____

4. Write the sums.

5 + 5 = _____

7 + 2 = _____

_____ = 1 + 4

_____ = 2 + 3

Use with Lesson 5.6.

How Much More? How Much Less?

Find each difference.

1. John Ⓟ Ⓟ Ⓟ Ⓟ Ⓟ Ⓟ Ⓟ Ⓟ

Nick Ⓟ Ⓟ

Who has more? _____ How much more? ____¢

---

2. June Ⓟ Ⓟ Ⓟ Ⓟ Ⓟ Ⓟ Ⓟ Ⓟ Ⓟ

Mia Ⓟ Ⓟ Ⓟ Ⓟ Ⓟ Ⓟ

Who has less? _____ How much less? ____¢

---

3. Carl 12 pennies

Mary 20 pennies

Who has more? _____ How much more? ____¢

Challenge

4. Dan 56 pennies

Bob 72 pennies

Who has less? _____ How much less? ____¢

Math Boxes 5.7

1. Solve the riddles.

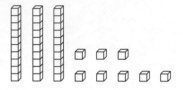

What am I? _____

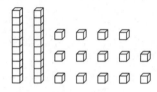

What am I? _____

2. Write a tally for 32.

Odd or even? _____

3. How many days since the 10th day of school?

_____

4. Write $<$, $>$, or $=$.

(N)(N) ☐ (D)

20¢ ☐ (N)(P)

24¢ ☐ $0.18

(D)(D)(D) ☐ 40¢

Use with Lesson 5.7.

Comparisons

Example

Mike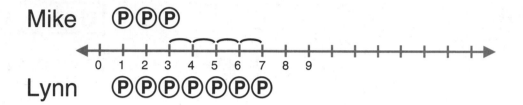

Lynn

Who has more? _____ How much more? ____

---

1. Amy

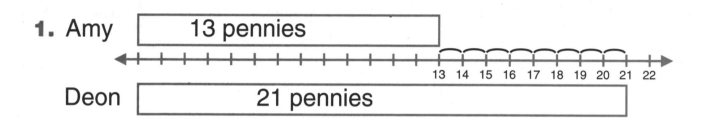

Deon

Who has more? _____ How much more? ____

---

2. Cat

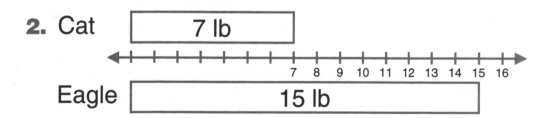

Eagle

Who weighs more? _____ How much more? ____

---

3. Andy | 17 chocolate kisses |

Kate | 25 chocolate kisses |

Who has less? _____ How much less? _____

Number Stories

Here is a number story Mandy made up.

I have 4 balloons.
Jamal brought 1 more.
We have 5 balloons
together.

$4 + 1 = 5$

Unit
balloons

Record your number story.

Unit

Math Boxes 5.8

1. Draw the hands.

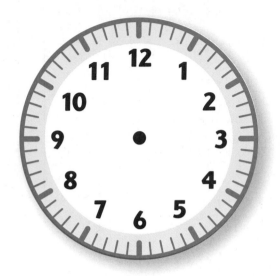

quarter-after 9 o'clock

2. Circle the tens place.

7 3

Is the number in the tens place odd or even?

_____

3.

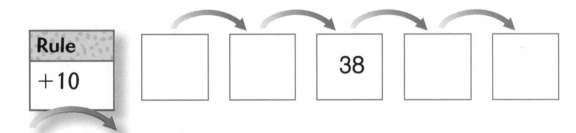

Rule
+10

38

4. Draw a line segment about 2 inches long.

Dice-Throw Record

Roll a pair of dice. Draw an X in a box for the sum, from the bottom up. Which number got to the top first? _____

2	3	4	5	6	7	8	9	10	11	12

Use with Lesson 5.9.

Differences with Base-10 Blocks

1. Nina ▯▯ooooo = ____

   Mario ▯▯ooooooooo = ____

   Who has more? _____ How much more? ____

2. Pam ▯▯▯▯ooooooo = ____

   Luis ▯▯▯▯▯ooooo = ____

   Who has more? _____ How much more? ____

3. Nate ▯▯▯▯▯ = ____

   Joy ▯▯▯ooooo = ____

   Who has more? _____ How much more? ____

4. Chaz ▯▯▯▯▯▯oo = ____

   Cori ▯▯ooooooo = ____

   Who has less? _____ How much less? ____

Math Boxes 5.9

1. Show 58¢ in two ways.
Use Ⓓ, Ⓝ, and Ⓟ.

2.

Lori | 14 pennies |

Beth | 25 pennies |

Who has more?

_____

How much more?

_____

3. Write the numbers.

What am I? ____

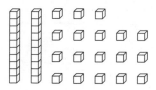

What am I? ____

4. Use your number grid.
Start at 90. Count back 25.

You end at _____.

$90 - 25 =$ _____

Turn-Around Facts Record

1 + 6 = ___	2 + 6 = ___	3 + 6 = ___	4 + 6 = ___	5 + 6 = ___	6 + 6 = ___
1 + 5 = ___	2 + 5 = ___	3 + 5 = ___	4 + 5 = ___	5 + 5 = ___	6 + 5 = ___
1 + 4 = ___	2 + 4 = ___	3 + 4 = ___	4 + 4 = ___	5 + 4 = ___	6 + 4 = ___
1 + 3 = ___	2 + 3 = ___	3 + 3 = ___	4 + 3 = ___	5 + 3 = ___	6 + 3 = ___
1 + 2 = ___	2 + 2 = ___	3 + 2 = ___	4 + 2 = ___	5 + 2 = ___	6 + 2 = ___
1 + 1 = ___	2 + 1 = ___	3 + 1 = ___	4 + 1 = ___	5 + 1 = ___	6 + 1 = ___

Use with Lesson 5.10.

Date

1. Complete the table.

Before	Number	After
13	14	15
	49	
	71	
	88	
	110	

2. Use | and . to show the number 42.

3. Record the time.

quarter-before _____ o'clock

4. Write $<$, $>$, or $=$.

7 + 6 ☐ 12

13 ☐ 6 + 7

14 − 6 ☐ 7

8 ☐ 15 − 8

Use with Lesson 5.10.

Facts Table

0 + 0 0	0 + 1 1	0 + 2 2	0 + 3 3	0 + 4 4	0 + 5 5	0 + 6 6	0 + 7 7	0 + 8 8	0 + 9 9
1 + 0 1	1 + 1 2	1 + 2 3	1 + 3 4	1 + 4 5	1 + 5 6	1 + 6 7	1 + 7 8	1 + 8 9	1 + 9 10
2 + 0 2	2 + 1 3	2 + 2 4	2 + 3 5	2 + 4 6	2 + 5 7	2 + 6 8	2 + 7 9	2 + 8 10	2 + 9 11
3 + 0 3	3 + 1 4	3 + 2 5	3 + 3 6	3 + 4 7	3 + 5 8	3 + 6 9	3 + 7 10	3 + 8 11	3 + 9 12
4 + 0 4	4 + 1 5	4 + 2 6	4 + 3 7	4 + 4 8	4 + 5 9	4 + 6 10	4 + 7 11	4 + 8 12	4 + 9 13
5 + 0 5	5 + 1 6	5 + 2 7	5 + 3 8	5 + 4 9	5 + 5 10	5 + 6 11	5 + 7 12	5 + 8 13	5 + 9 14
6 + 0 6	6 + 1 7	6 + 2 8	6 + 3 9	6 + 4 10	6 + 5 11	6 + 6 12	6 + 7 13	6 + 8 14	6 + 9 15
7 + 0 7	7 + 1 8	7 + 2 9	7 + 3 10	7 + 4 11	7 + 5 12	7 + 6 13	7 + 7 14	7 + 8 15	7 + 9 16
8 + 0 8	8 + 1 9	8 + 2 10	8 + 3 11	8 + 4 12	8 + 5 13	8 + 6 14	8 + 7 15	8 + 8 16	8 + 9 17
9 + 0 9	9 + 1 10	9 + 2 11	9 + 3 12	9 + 4 13	9 + 5 14	9 + 6 15	9 + 7 16	9 + 8 17	9 + 9 18

Easy Addition Facts

Complete.

Doubles Facts

$$0 + 0 = \boxed{}$$

$$6 + 6 = \boxed{}$$

$$1 + 1 = \boxed{}$$

$$7 + 7 = \boxed{}$$

$$2 + 2 = \boxed{}$$

$$8 + 8 = \boxed{}$$

$$3 + 3 = \boxed{}$$

$$9 + 9 = \boxed{}$$

$$4 + 4 = \boxed{}$$

$$10 + 10 = \boxed{}$$

$$5 + 5 = \boxed{}$$

10 Sums

$$0 + \boxed{} = 10$$

$$\boxed{} + 4 = 10$$

$$\boxed{} + 9 = 10$$

$$7 + \boxed{} = 10$$

$$\boxed{} + 8 = 10$$

$$\boxed{} + 2 = 10$$

$$3 + \boxed{} = 10$$

$$\boxed{} + 1 = 10$$

$$\boxed{} + 6 = 10$$

$$10 + \boxed{} = 10$$

$$\boxed{} + 5 = 10$$

Use with Lesson 5.11.

Measurement Practice

1. Name or draw an object. Estimate its length to the nearest inch. Measure it with a ruler to the nearest inch.

Object (Name it or draw it)	My estimate	My measurement
	about _____ inches	about _____ inches
	about _____ inches	about _____ inches
	about _____ inches	about _____ inches

Measure these line segments to the nearest inch.

2. _____   about ____ in.

3. _____   about ____ in.

4. _____   about ____ in.

5. Draw a line segment 4 inches long.

Challenge

6. Measure to the nearest half-inch.

_____   about _____ in.

7. Draw a line segment 5 and a half inches long.

Math Boxes 5.11

1. Roger   Ⓓ Ⓟ Ⓟ Ⓓ Ⓓ

    Lauren   Ⓓ Ⓟ Ⓟ Ⓝ Ⓟ Ⓝ Ⓝ

Who has more?

_____

How much more?

_____

2. Write the numbers.

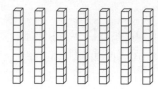

What am I? _____

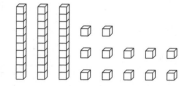

What am I? _____

3. Fill in the dots and numbers.

$9 = 7 + \underline{\quad}$ $\underline{\quad} + 4 = 8$

   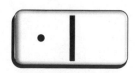

$3 + \underline{\quad} = 10$ $7 = 1 + \underline{\quad}$

4.   **Judy's Dice Rolls**

				×	
×				×	
×		×		×	
×		×	×	×	×
×	×	×	×	×	×
1	2	3	4	5	6

How many times did Judy

roll?_____

What number came up the

most? _____

     Use with Lesson 5.11.

"What's My Rule?"

Find the missing numbers and rules.

1.

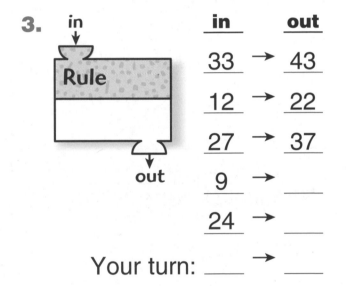

in		out
7	→	4
11	→	8
4	→	1
9	→	___

Your turn: ___ → ___

2.

in		out
5	→	10
8	→	13
12	→	17
16	→	___

Your turn: ___ → ___

3.

in		out
33	→	43
12	→	22
27	→	37
9	→	___
24	→	___

Your turn: ___ → ___

4. Challenge

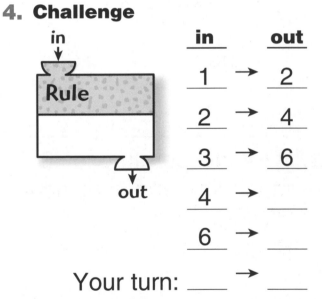

in		out
1	→	2
2	→	4
3	→	6
4	→	___
6	→	___

Your turn: ___ → ___

Make your own.

5.

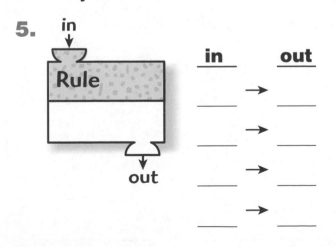

in		out
___	→	___
___	→	___
___	→	___
___	→	___

6.

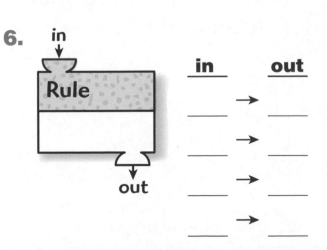

in		out
___	→	___
___	→	___
___	→	___
___	→	___

1. Draw the hands.

quarter-past 8 o'clock

2. Find something about 2 inches long. Draw a picture of what you found.

3. If the temperature is 42°F in the A.M. and goes up to 68°F at noon, how many degrees warmer is it?

_____

(*Hint:* You may use your number grid.)

4.

Reggie | 31 pennies |

Sam | 22 pennies |

Who has more?

_____

How much more?

_____

"What's My Rule?"

Find the rule.

1.

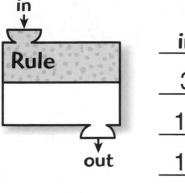

in	out
3	5
12	14
10	12

Your turn: _____

2.

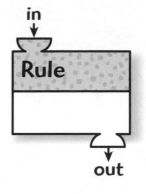

in	out
4	1
12	9
17	14

Your turn: _____

What comes out?

3.

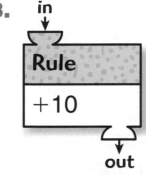

in	out
3	13
16	
25	

Your turn: _____

4.

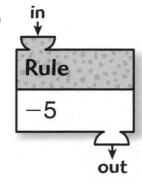

in	out
10	5
8	
15	

Your turn: _____

5. What goes in?

Rule
−3

in	out
	5
	3
	2

Your turn: _____

6. Make your own.

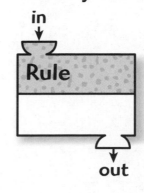

in	out

Math Boxes 5.13

1. Add.

$4 + 4 =$ _____

_____ $= 0 + 9$

$\begin{array}{r} 3 \\ +\ 2 \\ \hline \end{array}$ $\begin{array}{r} 8 \\ +\ 1 \\ \hline \end{array}$

2. Write $<$, $>$, or $=$.

37¢ ☐ $0.35

$2 + 6$ ☐ 9

Ⓓ Ⓓ Ⓟ ☐ $0.30

82 ☐ 28

3. Write the numbers.

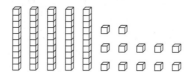

What am I? _____

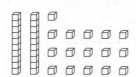

What am I? _____

4. Find the missing rule. Fill in the missing numbers.

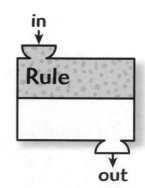

in	out
3	5
17	19
14	16
10	
25	

Use with Lesson 5.13.

Math Boxes 5.14

1. Use ❘ and ▪ to show 53 in two ways.

2. Lois ⓅⓅⒹⓃⒹⓃ

Joe ⒹⓅⓅⓅⓅⓃ

Who has more?

_____

How much more?

_____

3.

A 19-pound  and a

56-pound weigh

_____ pounds together.
(*Hint:* You may use your number grid.)

4. Use <, >, or =.

Equivalents and Abbreviations Table

Weight

kilogram: 1,000 g
pound: 16 oz
ton: 2,000 lb
1 ounce is about 30 g

<	is less than
>	is more than
=	is equal to
=	is the same as

Length

kilometer: 1,000 m
meter: 100 cm or
10 dm

foot: 12 in.
yard: 3 ft or 36 in.
mile: 5,280 ft or
1,760 yd

10 cm is about 4 in.

Time

year:	365 or 366 days
year:	about 52 weeks
year:	12 months
month:	28, 29, 30, or 31 days
week:	7 days
day:	24 hours
hour:	60 minutes
minute:	60 seconds

Money

 1¢, or $0.01   Ⓟ

 5¢, or $0.05   Ⓝ

 10¢, or $0.10   Ⓓ

 25¢, or $0.25   Ⓠ

 100¢, or $1.00 $1

Abbreviations

kilometers	km
meters	m
centimeters	cm
miles	mi
feet	ft
yards	yd
inches	in.
tons	t
pounds	lb
ounces	oz
kilograms	kg
grams	g

Date

Notes

Date

Notes

Date

Notes

Date

Notes

Notes

Notes

Number Cards 0-15

15	**14**	**13**	**12**
11	**10**	**9**	**8**
7	**6**	**5**	**4**
3	**2**	**1**	**0**

Number Cards 16–22

16	17	18	19
20	21	22	+
−	×	÷	=
<	?	wild card	wild card

Number Cards 0-9

1   0

5   4   3   2

9   8   7   6

Domino Cutouts

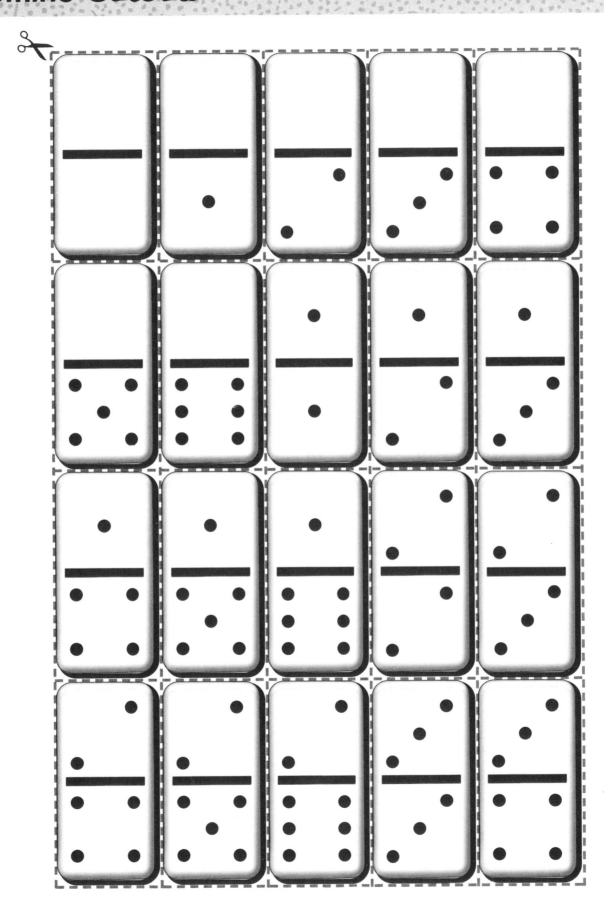

Domino Cutouts

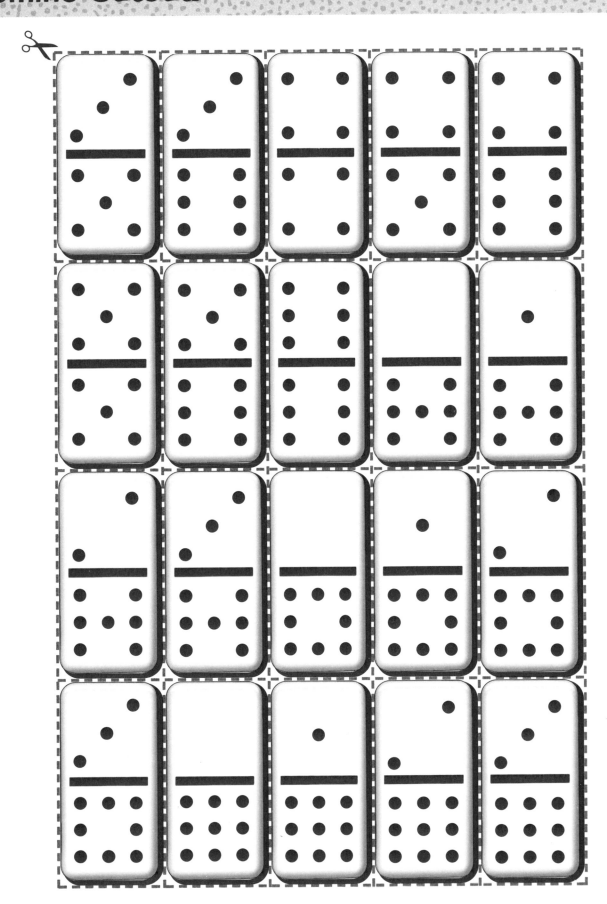

Activity Sheet 4

Name

Date

Base-10 Pieces

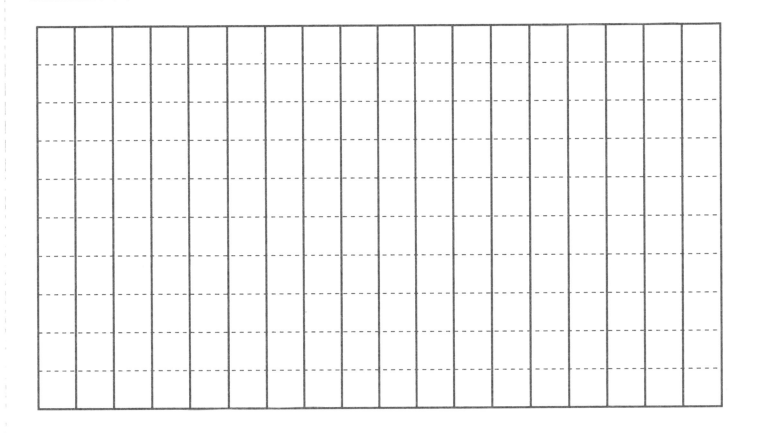

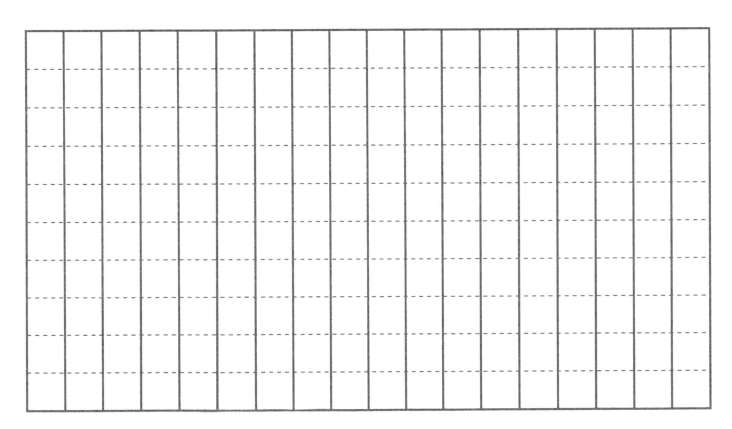

Base-10 Pieces

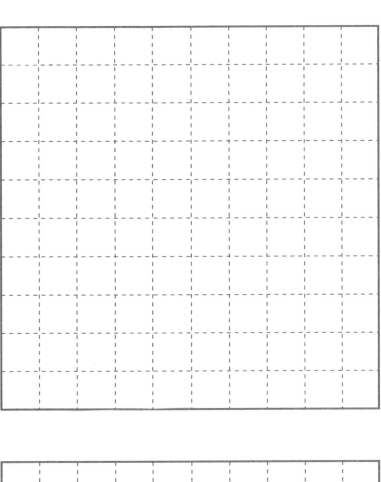

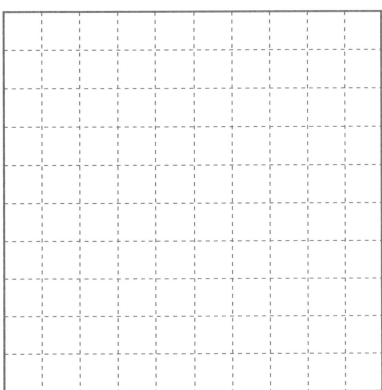

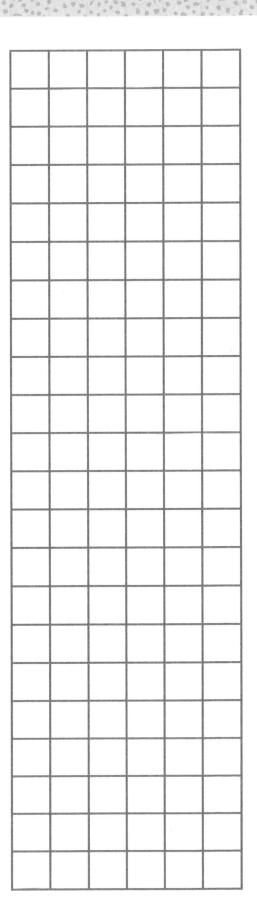

Activity Sheet 6

Animal Cards

first-grade girl
41 lb

8-year-old boy
50 lb

cheetah
120 lb

porpoise
98 lb

penguin
75 lb

beaver
56 lb

Use with Lesson 5.5.

Activity Sheet 7

Animal Cards

8-year-old boy
60 in.

first-grade girl
43 in.

porpoise
72 in.

cheetah
48 in.

beaver
30 in.

penguin
36 in.

Animal Cards

cat
7 lb

fox
14 lb

koala
19 lb

raccoon
23 lb

rabbit
6 lb

eagle
15 lb

Animal Cards

fox

20 in.

cat

12 in.

raccoon

23 in.

koala

24 in.

eagle

35 in.

rabbit

11 in.